YOUR KNOWLEDGE HAS VALUE

- We will publish your bachelor's and
  master's thesis, essays and papers

- Your own eBook and book -
  sold worldwide in all relevant shops

- Earn money with each sale

Upload your text at www.GRIN.com
and publish for free

Bibliographic information published by the German National Library:

The German National Library lists this publication in the National Bibliography;
detailed bibliographic data are available on the Internet at http://dnb.dnb.de .

This book is copyright material and must not be copied, reproduced, transferred, distributed, leased, licensed or publicly performed or used in any way except as specifically permitted in writing by the publishers, as allowed under the terms and conditions under which it was purchased or as strictly permitted by applicable copyright law. Any unauthorized distribution or use of this text may be a direct infringement of the author s and publisher s rights and those responsible may be liable in law accordingly.

Imprint:

Copyright © 2018 GRIN Verlag
Print and binding: Books on Demand GmbH, Norderstedt Germany
ISBN: 9783668752610

This book at GRIN:

https://www.grin.com/document/429538

Mutinda Jackson

The different ways in which viruses cause tumors

The Epstein-Barr, Hepatitis B and Human papilloma virus

GRIN Verlag

GRIN - Your knowledge has value

Since its foundation in 1998, GRIN has specialized in publishing academic texts by students, college teachers and other academics as e-book and printed book. The website www.grin.com is an ideal platform for presenting term papers, final papers, scientific essays, dissertations and specialist books.

Visit us on the internet:

http://www.grin.com/

http://www.facebook.com/grincom

http://www.twitter.com/grin_com

Viruses and the Way they Cause Tumours

Word Count: 2,573

Introduction

The burden associated with viral infections in cancer remains to be high, although it has been overlooked by a number of cancer researchers. However, several studies on viruses have cropped up from the cancer field, providing significant insights into both non-infectious and infectious causes of cancer. Since the avian cancer virus experiment of the year 2011, scientists have confirmed seven different viruses that cause about ten to fifteen percent of human cancer globally. Notably, these distinct viruses have revealed unprecedented links between innate immunities as well as sensors, and tumour suppressor, implying that they control both cancer and viral infections.

Apparently, these varied cancers remain to be a substantial health challenge, especially in developing economies, alongside the underserved and immuno-suppressed individuals in developed economies (Cancer Research UK, 2014). Again, these cancers are comprised of readily identifiable diagnosis, prevention and therapy targets. Among the several human tumour viruses evidenced by several scientists and scholars, this paper will explore the Epstein - Barr virus (EBV), Hepatitis B virus (HBV) and Human papilloma virus (HPV) along with their distinct ways by which they cause tumours.

Epstein-Barr virus (EBV)

EBV Microbe Description

The EBV, which is also termed as the human herpes virus 4 (HHV4) was described fifty-three years after the initial experiments of Rous; in 1964 in the identification of the EBV particles in cell lines from Burkitt's lymphoma African patients (Moore & Chang, 2010). The EBV is a herpes virus that contains a large double-stranded DNA genome, and just like all other herpesviruses, it encodes enzymes that are involved in DNA replication as well as repair and nucleotide biosynthesis. It also possesses the ability to establish latency in B lymphocytes, reactivating into a lytic cycle (Liao, 2006). In other words, EBV infects and stays in certain WBCs (white blood cells) in the body known as B lymphocytes (B cells), and it is renowned in causing infectious mononucleosis or naturally occurring human tumours.

EBV Disease Explanation

According to scientists, EBV remains to be an ever-present virus, which is often common recognized as the principal agent for infectious mononucleosis, mostly termed as 'mono' or the 'kissing disease.' Research demonstrates that about ninety-five percent of all adults have been estimated to be seropositive, with most of the EBV infections being subclinical (Chiras, 2013). In addition, EBV has been linked with several malignancies including B and T cell lymphomas, leiomyosarcomas, Hodgkin's disease, post-transplant lymphoproliferative disease alongside nasopharyngeal carcinomas. Among these cancers, Burkitt's lymphoma, leiomyosarcomas and post-transplant lymphoproliferative diseases have been noted to demonstrate an increased frequency in patients who have immunodeficiency; hence, suggesting immunosurveillance role in the malignant transformation suppression (Moore & Chang, 2010).

Course of the EBV Infection

The oropharyngeal cavity remains to be the primary site of the EBV, and the virus has the ability of infecting the B cells and the epithelial cells, alongside switching between these two cells. Apparently, the major surface glycoprotein-gp 350/220, attaches itself to the cd21 receptor cells on the B cells (Llovet *et al.* 2003). The B cells' transformation remains to be a highly efficient process that requires a large portion of the genome of the EBV, which develops into circular for both replication and latency (Liao, 2006). As a result, the virus will directly enter the latent gene expression state with the lytic cycle suppression.

Current Research and Future Research/Directions

Significantly, immune therapy of tumours linked with EBV has currently been the focus of research because the standard therapy has commonly involved the utilization of radiation therapy, multi-agent chemotherapy and together with surgery (Lavanchy, 2004). It is a phenomenon that has concentrated on adoptive transfer of specific cytotoxic T-cells of EBV and exhibited significant success, though it has to overcome some obstacles including potential graft versus host disease and resistance because of selected EBV epitopes' mutation (Moore & Chang, 2010).

Currently, vaccines that have the ability of preventing primary infection of the EBV or that can boost immune responses against tumours related with EBV remain under investigation. So far, much of the development has focused mainly on gp 350/220 subunit vaccines because it represents one of the most abundant proteins on the coat of the virus, on top of being the protein against which the neutralizing antibody response of the human EBV is directed (Moore & Chang, 2010). Another direction entails the use of a recombinant vaccinia viral vector in expressing the membrane antigen of an EVB since a successful vaccine is supposed to have the highest impact in global regions that have high incidences of certain malignancies (Liao, 2006).

World Relevance

As aforementioned, since a successful vaccine has to possess the highest impact in global regions that have high incidences of certain malignancies, directions that involve the use of a recombinant vaccinia viral vector so as to express an EVB antigen have to be stressed. For instance, Burkitt's lymphoma has been observed to be the highest common childhood malignancy in the African continent, especially in central Africa where EBV and malaria are measured as cofactors in its carcinogenesis (Chiras, 2013). Recent researches have demonstrated that about ninety-five percent of children in these regions are affected by the age of three, unlike in the United States and other places where infection is normally delayed till adolescence. Again, nasopharyngeal carcinoma has been observed to be rather rare, although it has an unusually high frequency in southern China, close to more than twenty times higher compared to that of the most populations (Moore & Chang, 2010).

Hepatitis B virus (HBV)

HBV Microbe Description

Hepatitis B virus, unlike Hepatitis C virus, is a DNA virus of the hepadnaviridae family, although the features of the resulting diseases of the two share several similarities. The HBV virion is a 42-nm particle that has an electron-dense core (nucleocapsid) of a diameter of 27 nm, enveloped by the surface protein (HBsAg) that is entrenched in a membranous lipid copied from the cell of the host. The virion genome of the HBV is circular and has about 3.2 kb in size, consisting of DNA that is mostly double-stranded (Barreto *et al*. 2014). Moreover, it has a compact organization containing four overlapping reading frames in the same direction and has no noncoding regions. The minus strand is unit length and consists of a protein that is covalently joined to the 5' end while the other strand (plus strand) is variable in length, although it contains not more than unit length, besides having an RNA oligonucleotide at its 5' end (Moore & Chang, 2010). Therefore, there is no strand that is closed, and circularity is usually maintained by cohesive ends. The HBV's four overlapping open reading frames in the genome aid the transcription and expression of varied hepatitis B proteins, a process that takes place through the use of the manifold in-frame start codons (Chiras, 2013). Again, the HBV genome contains parts responsible for the regulation of the transcription, determination of the polyadenylation site and regulation of a specific transcript nucleocapsid encapsidation.

HBV Disease Explanation

Remarkably, hepatitis B is a blood-borne pathogen, which can cause acute and chronic hepatitis. In this sense, chronic hepatitis; infections lasting for more than three months, have the potential of causing cirrhosis and liver failure (Liao, 2006). Most importantly, chronic infections can result in the establishment of hepatocellular carcinoma. Scientists affirm that the hepatocellular carcinoma is an insistent tumour that may occur in the liver disease setting, arising from infections with HBV, though the exact oncogenesis mechanism with this virus remains uncertain (Chiras, 2013). Acute HBV infection only causes mild symptoms, with a significant number of infected adults successfully clearing the virus and acquiring a lasting immunity.

HBV Disease Course

The surface antigen appears in most patients' sera in the time of incubation, two to eight weeks before the onset of jaundice or the biochemical liver damage evidence. The antigen persists in the acute illness, normally clearing from the circulation in times of convalescence. The DNA polymerase activity, which is linked with the virus, appears in the circulation, correlating in time with liver cells damage, facilitated by increased serum transaminases (Moore & Chang, 2010). The polymerase activity will last for some days or weeks in acute situations, and for several months or years in certain persistent carriers (Barreto *et al*. 2014). The antibody to the core antigen remains in the serum two to ten weeks after the appearance of the surface antigen, and it is often noticeable for several years after recovery. Afterward, the core antibody titer appears so as to correlate with the virus replication amount and duration, as eventually, the antibody to the surface antigen component materializes (Liao, 2006).

Interaction between HBV Microbe and Host

The antibody together with the cell-mediated immune responses to diverse types of antigens has been observed to be induced in the time of infection. Nevertheless, this does not always tend to be protective, and in some cases, can lead to an autoimmune phenomenon that contributes to disease pathogenesis (Yount, 2009). The immune response to HBV infection is directed to several antigens including hepatitis B surface antigen and the core antigen (Hoffman, 2007). The presupposition that hepatitis B exerts its detrimental effect on hepatocytes through direct cytopathic changes remains inconsistent with the large quantities persistence of the surface antigen in liver cells of several healthy persons who are carriers.

HBV Current Research and Future Research/Directions

Although the traditional basis of treatment remains to be surgical, whether in the resection of tumour or plantation, nonsurgical options including radiofrequency ablation, percutaneous ethanol injection, chemotherapy, transarterial embolization and even radiotherapy are also used. The selection of these copious therapies usually depends on the extent of the illness along with the amount of the liver function that the patient has in reserve (Peralta-Zaragoza *et al*. 2013). New therapies researchers have emphasized on the utilization virally targeted as well as immunological strategies with the intent of preventing infection. HBV vaccine introduction in the early 1980s marked a major landmark for cancer prevention vaccine; more than one-hundred nations worldwide have adopted a universal policy of newborn immunization since then (Zheng & Baker, 2006). Nonetheless, due to the precincts of the existing vaccines, multiple dosing schedule is being attended to with efforts of merged it with other requisite vaccines or decreasing the number of doses (Lavanchy, 2004).

Global Significance

The realization of the variation in the epitopes found on the virions surface and in the subviral particles, helped in identifying numerous HBV subtypes that vary in their geographical distribution (Hubert & Laimins, 2002). Therefore, these variations ease the application of vaccines in the various regions as per the HBV predominance. Additionally, in other regions with HBsAg carriage prevalence like China and Southeast Asia, the transmission predominance route remains perinatal; thus, encouraging the administration of a vaccine course immediately after birth so as to effectively prevent transmission from a positive mother (Moore & Chang, 2010).

Human Papilloma Virus (HPV)

HPV Microbe Description

HVP represents a group of small non-enveloped viruses of a DNA tumour, and they contain a virion size of about 55 nm in diameter. HVPs normally cause benign tumour, although sometimes they can also cause malignancies (Spink & Laimins, 2005). The viral genome is comprised of a circular double-stranded DNA, approximately 8 kb in size, renowned in encoding for early proteins such as E1, E2, E4, E5, E6 and E7, which contribute to virus replication alongside cell transformation, and for late L1 and L2 proteins-structural units of the viral capsid (Peralta-Zaragoza *et al.* 2013).

Description of HBV Infection

Evidently, HPV is widely recognized to cause benign papillomas or the so-termed as warts in human beings. Persistent infection with the human papillomavirus subtypes that are highly risky has been broadly linked with the development of cervical cancer (Yount, 2009). The HPV has been observed to infect the epithelial cells, and integration in host DNA, the production of the oncoproteins, especially E6 and E7, the virus disrupts the natural pathways of the tumour suppressor and is needed for cervical carcinoma cells proliferation (Frazer, 2004). In addition, this virus highly contributes to the development of other human cancers, for example, neck and head tumours, and skin cancers among immunosuppressed patients alongside other anogenital cancers.

The Course of HPV Infection

Notably, the immune system plays a significant role in the prevention of persistent HPV infection as well as the progression of precancerous lesions. According to scientific evidence, the human papillomavirus remains to be a poor natural immunogen, and since it is a double-stranded DNA virus, it allows the initiation of innate responses due to lack RNA intermediate and the inability of the infection to cause cytolysis (Schiff, 2006). Mainly, the HPV encodes nucleoproteins that are not secreted, and which are poorly cross-presented and compared to different viruses that its non-structural proteins are expressed, mainly at low levels. Conversely, HPV genital infection remains transient. In addition, inadequate responses of the T-cells may result in failure to clear cells that have been infected by the HPV (Moore & Chang, 2010)..

Current Research and Future Research/Directions

Based on clinical trials, which showed about one-hundred percent, protection from persistent infection via high level neutralizing antibodies generation, the FDA approved the use of an effective prophylactic vaccine to fight HPV 16 and 18 based on virus-like L1 recombinant particles (the major capsid protein) in 2006 (Liao, 2006). Thus, since such types remain to be the causative agent of about seventy percent of cervical cancers, the development of this effective vaccine has a substantial promise for cervical cancer prevention. Conversely, the vaccine, which is extremely expensive, does not provide protection against other HPV types that may also be highly risk; hence, will probably have limited advantage to women who are already infected (Moore & Chang, 2010).

As a result of these limitations, therapeutic vaccinations are undergoing research so as to come up with new strategies of treating women who are already infected, alongside accelerating the effect of prophylactic vaccination so as to decrease cervical cancer occurrences (Zheng & Baker, 2006).

Global Significance

Cervical cancer has been considered to be the second major cause of cancer mortality among women in the globe, causing more than two-hundred and forty fatalities per annum (Hoffman, 2007). Of about four-hundred and ninety thousand cases that are reported each year, more than eighty percent have been defined to occur in underdeveloped nations, where effective, though costly screening programs like Pap smear are not in place (Schiff, 2006). Early precancerous changes and cancers detected using Pap smears have been noted to be effectively treated and cured with either ablation or surgical therapy. Thus, this situation encourages the development of effective screening in underdeveloped economies because its absence denotes a late detection of the disease.

Conclusion

The viruses explored above have illustrated the distinct biological malignancy pathways, alongside their respective resulting tumours. However, the viral gene products present in cancer and even precancerous cells present significant targets, which can be exploited in current and future therapies that differentiate these cells from the normal ones. Therefore, targeting cancer cells will have substantial benefits compared to traditional modalities including radiotherapy and chemotherapy. For example, cervical cancer can provide an ideal model for cytotoxic immune therapies as it retains the HPV viral oncoproteins E6 and E7, alongside requiring their persistent proliferation expression. With the prevalence of the aforementioned cancers in developing countries, together with health care infrastructure limitations, the novel vaccine strategies principally designed to prevent infection as well as targeted therapies for the disease treatment, have to be carefully considered in this milieu.

References

Barreto, S. C., Uppalapati, M., & Ray, A. 2014. Small Circular DNAs in Human Pathology. The Malaysian Journal of Medical Sciences: *MJMS*, 21(3), 4–18.

Cancer Research UK. 2014. *What causes cancer?* Viewed from http://www.cancerresearchuk.org/about-cancer/cancers-in-general/causes-symptoms/causes/what-causes-cancer

Chiras, D. 2013. *Human Biology*. Boston: Jones & Bartlett Publishers.

Frazer, I. H. 2004. Prevention of cervical cancer through papillomavirus vaccination. *Nat Rev Immunol*, 4(*1*):46–54.

Hoffman, E. J. 2007. *Cancer and the Search for Selective Biochemical Inhibitors, Second Edition*. Florida: CRC Press.

Hubert, W. G. & Laimins, L. A. 2002. Human papillomavirus type 31 replication modes during the early phases of the viral life cycle depend on transcriptional and posttranscriptional regulation of E1 and E2 expression: *Journal of Virology*, 76:2263–2273

Lavanchy, D. 2004. Hepatitis B virus epidemiology, disease burden, treatment, and current and emerging prevention and control measures: *Journal of Viral Hepatitis*, 11(*2*): 97–107.

Liao, J. B. 2006. Viruses and Human Cancer: Yale Journal of Biology and Medicine, 79(2-3): 115-122.

Llovet, J. M, Burroughs, A. & Bruix, J. 2003. Hepatocellular carcinoma: *Lancet*, 362(*9399*):1907–1917.

Moore, P. & Chang, Y. 2010. Why do viruses cause cancer? Highlights of the first century of human tumour virology: *Nat Rev Cancer*, 10(*12*): 878-89.

Peralta-Zaragoza, O., Deas, J., Gomez-Ceron, C., Garcia-Suastegui, W. A., Fierros-Zarate, G. del S., & Jacobo-Herrera, N. J. 2013. HPV-Based Screening, Triage, Treatment, and Followup Strategies in the Management of Cervical Intraepithelial Neoplasia. *Obstetrics and Gynecology International*, 2013, 912780. doi:10.1155/2013/912780

Schiff, E. R. 2006. Prevention of mortality from hepatitis B and hepatitis C. *Lancet*, 368(*9539*):896–897.

Spink, K. M & Laimins, L. A. 2005. Induction of the human papillomavirus type 31 late promoter requires differentiation but not DNA amplification: *Journal of Virology*, 79:4918–4926.

Yount, L. 2009. *A to Z of Biologists*. New York: Infobase Publishing.

Zheng, Z. & Baker, C. 2006. Papillomavirus Genome Structure, Expression, and Post-Transcriptional Regulation: *Journal of Virtual Library*, 11: 2286-2302.

YOUR KNOWLEDGE HAS VALUE

- We will publish your bachelor's and master's thesis, essays and papers

- Your own eBook and book - sold worldwide in all relevant shops

- Earn money with each sale

Upload your text at www.GRIN.com and publish for free